3- MINUTE STEPHEN HAWKING

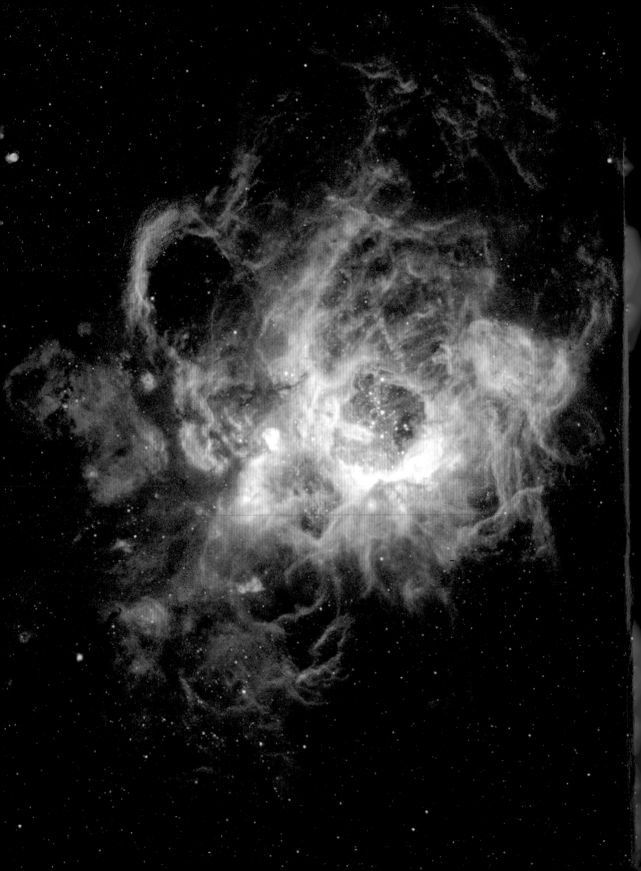

3-MINUTE
STEPHEN
HAWKING

His life, theories, and influence
in 3-minute particles

Paul Parsons & Gail Dixon

Foreword by **John Gribbin**

METRO BOOKS
NEW YORK

METRO BOOKS
New York

An Imprint of Sterling Publishing
387 Park Avenue South
New York, NY 10016

This 2012 edition published by Metro Books
by arrangement with **Ivy Press**

This book was conceived,
designed, and produced by
Ivy Press
210 High Street, Lewes,
East Sussex BN7 2NS, UK
www.ivy-group.co.uk

CREATIVE DIRECTOR   Peter Bridgewater
PUBLISHER   Jason Hook
EDITORIAL DIRECTOR   Caroline Earle
ART DIRECTOR   Michael Whitehead
DESIGNER   Glyn Bridgewater
PROJECT EDITOR   Jamie Pumfrey

ISBN 978-1-4351-4064-6

Manufactured in China

2 4 6 8 10 9 7 5 3 1

www.sterlingpublishing.com

Contents

Chapter 2: Theories 64

Chapter 3: Influence 110

Foreword

Like Albert Einstein, Stephen Hawking has become an iconic symbol of science, known to millions of people who have little or no understanding of the work he does. The image of a brilliant mind trapped in a failing body has helped to encourage a fascination with the man, while the subjects of his study—black holes, the birth of the Universe, and the end of time—are fascinating in themselves. It sometimes seems to me that people take an almost perverse delight in not understanding these subjects, as if it would somehow cheapen them, or remove the magic, to have the work of people such as Hawking explained in simple terms. When his book *A Brief History of Time* was first published, and famously described as the least-read bestseller of all time, I heard people at dinner parties competing to display their ignorance of its subject matter. If one person said, "Well, I couldn't get beyond chapter 2," another was sure to chime in along the lines of, "I couldn't get past page 2!"

This is a pity, because the mystery and magic of the kind of science that Stephen Hawking has been involved with is actually enhanced, not diminished, by a little understanding. It is Paul Parsons and Gail Dixon that provide enhancement here. The story of Hawking's life is well known, at least in outline, and everyone can empathize with and admire the way in which he has overcome his physical difficulties. But to me, the far more significant story is the way in which Hawking has contributed to the dramatic development in our understanding of the Universe that took place in the second half of the twentieth century. I first encountered Hawking when I was a student visiting Cambridge in 1967, and he took part in a debate about the Big Bang. At that time, the Big Bang theory was still not fully recognized as the best description of the Universe around us. Today, it is not only accepted but the date of the Big Bang has been pinned down to 13.7 billion years ago (not 13.6 or 13.8!), and researchers such as Hawking have developed ideas about the kind of event that must have triggered the birth of the Universe as we know it.

This is the grand theme of Hawking's work, which Paul and Gail explain concisely and clearly in these pages, giving everyone the opportunity to share the magic of science at the frontiers of knowledge. Anyone who has ever given up on one of Hawking's own books after chapter 2 (or page 2!) should read this book about the man and his work and try again; if you do not, you will be missing a treat.

JOHN GRIBBIN
Visiting Fellow in Astronomy
University of Sussex, UK

How the Book Works

Stephen Hawking is regarded by many as the greatest physicist alive today—a living genius. This book tells his story in three parts. The first deals with his incredible life. Diagnosed with a crippling medical condition at the age of 21, he is now confined to a wheelchair and only able to talk using an electronic speech synthesizer. His seminal contributions to physics and cosmology are set out in lay-language in chapter two. Hawking has shown how black holes play host to infinitely dense "singularities" at their cores—and that the Universe itself may have begun in a similar state. He famously proved that black holes aren't so black but can actually emit particles and radiation. And he later produced theories blending relativity and quantum mechanics to explain how our Universe may have been created from nothing. Finally, chapter 3 takes stock of Hawking's influence: how he has not only set the bar for modern research in quantum physics and relativity but also become a campaigner for numerous important causes.

Chapter 1
Life

Chapter 2
Theories

Chapter 3
Influence

3-minute Stephen Hawking

Each chapter of the book breaks down into 20 short sections. The Theories chapter, for example, carries sections on black holes, Hawking radiation, and quantum cosmology—as well as a great many more. Each section splits into three paragraphs, and each of these deals with one aspect of that particular topic. For example, the section on black holes breaks down into "heart of darkness" (which explains what exactly a black hole is), "bad trip" (detailing what happens to anything unfortunate enough to fall into one), and "egg head" (examining the odd "no hair" property of black holes, which Hawking helped to prove mathematically). Each paragraph will take you about one minute to read, meaning that each section will take roughly three minutes. Hence, 3-Minute Stephen Hawking...

Chapter 2 **Theories**

Black Holes

Heart of darkness

Black holes are regions in space and time where gravity is so intense that not even light can escape. As far back as the eighteenth century, the English philosopher **John Michell** first mooted that such regions might exist. He called them "dark stars." But it wasn't until 1915 that German physicist **Karl Schwarzschild** showed how their existence was predicted by Einstein's general theory of relativity. Yet physicists didn't begin to explore the relationship between relativity and black holes fully until the 1940s. Indeed, the name "black hole" was only coined by Princeton University physicist **John Archibald Wheeler** in 1967.

Bad trip

A black hole's outer surface is known as its "**event horizon.**" It's not a solid surface at all, rather an imaginary sphere in space, centered on the black hole's core, from within which it is **impossible to escape** the black hole's gravitational pull. Anything falling over the event horizon can never escape and is doomed to be crushed out of existence at the black hole's center. In his books and lectures, Hawking is fond of describing what it might be like for a traveler to fall over a black hole's event horizon—using the word "**spaghettification**" as a rather gruesome term for the stretching effect of intense gravitational fields on the fragile human body.

Egg head

Black holes figured prominently in the Golden Age of research into the general theory of relativity, and Hawking contributed many new results. One of his early pieces of research was to prove, in collaboration with the Australian physicist **Brandon Carter**, that black holes have "no hair." This was an idea that had been put forward without proof by John Archibald Wheeler, who believed that a black hole could be specified by just three numbers—its mass, its electrical charge, and its rate of rotation. Hawking and Carter proved that this is indeed the case—a black hole's **powerful gravity** erases all other fine details, or "**hair**," so that two black holes with identical mass, charge, and spin would be—at least according to classical physics—indistinguishable.

3-second Brief
Hawking gets to work building a new understanding of the weirdest objects in the Universe—proving that, if they exist, they are actually some of the simplest objects in physics.

Related Thoughts
see also
SINGULARITIES
page 72
HAWKING RADIATION
page 80

"One could well say of the event horizon what the poet Dante said of the entrance to Hell: 'All hope abandon, ye who enter here.'"

Instant expert

By the same reckoning, each chapter will take you around an hour—and the whole book will guide you on a journey from Hawking's birth, through his golden age of discovery in the 1970s, and up to the present day, in about three hours. To aid understanding further, each chapter ends with a glossary of common terms encountered and a timeline that sets out clearly the order in which key events took place. It's the definitive quick-reference guide to perhaps the greatest, and most courageous, living physicist of our age.

Introduction

Stephen Hawking is one of the most eminent cosmologists of our time. Alongside Sir Isaac Newton and Einstein, he is among the greatest thinkers in theoretical physics of all time. His groundbreaking work has broadened our understanding of the Universe and helped to unravel its many mysteries. Hawking's academic career has spanned more than 50 years, and for most of that time he has had to battle with Amyotrophic Lateral Sclerosis (ALS), a disease that has left him almost completely paralyzed. He was given the dreadful diagnosis at the age of 21—and told he had just two years to live. But he never allowed himself to succumb to despair or self-pity. Instead, he found the courage and optimism to continue his research, rising to become Lucasian Professor of Mathematics at Cambridge University, a post he held for 30 years, and becoming one of the youngest ever Fellows of the prestigious Royal Society, Britain's premier scientific organization. His seminal popular-science book *A Brief History of Time* was a literary phenomenon. Published in 1988, it became an overnight success and stayed in the bestseller charts for four years. The book brought cosmology to the masses and created a newfound thirst for popular science that has resulted in a steady stream of books and television programs. *A Brief History of Time* made Hawking a superstar, and he has traveled the world imparting his vision of the cosmos in lectures that have inspired a generation of listeners. He is also an active campaigner for the rights of disabled people and has used his public profile to speak out against nuclear weapons, education cuts, and war in the Middle East. There is much to admire about Stephen Hawking—not least his staggering triumph over physical adversity—but it is his intellectual achievements that will stand the test of time. He has been a true successor to Einstein.

Stellar start

Stephen William Hawking was born on January 8, 1942, on the 300th anniversary of the death of Galileo. His parents Frank (above) and Isobel were both Oxford-educated intellectuals.

Early influence

Hawking's family mixed with the intelligentsia. Hawking (left) pictured with writer Robert Graves' son, William Graves, on holiday in Deya, Majorca, in 1951.

Teenage boffin

Stephen at his St. Albans home. A penetrating stare from Hawking, aged 12, says so much about his early promise. His genius had already earned him the nickname "Einstein."

Traditional times

Joining societies such as the Cub Scouts allowed Hawking some respite from the intense nature of his studies at school and large amounts of homework.

The graduate

Hawking found his studies at Oxford University undemanding and slid into a life of apathy. As a result, he barely managed to achieve the first-class honors he needed to get into Cambridge.

Against all odds

Despite being diagnosed with Amyotrophic Lateral Sclerosis (ALS) at the age of 21, and being given two years to live, Hawking went on to marry Jane Wilde and have a family.

Iconic voice

Since losing the power of speech, computer technology has allowed Hawking to communicate his ideas to the world—ideas that would otherwise remain locked in his brain.

Singular minds

Hawking with Oxford mathematician Roger Penrose (right), with whom he proved the existence of singularities—points of infinite density at the hearts of black holes. It was an idea Hawking would later extend to explain the birth of the Universe.

Canadian club

Hawking has collaborated with many of the world's leading scientists, including Neil Turok, director of the Perimeter Institute of Theoretical Physics at Ontario, Canada. Hawking has visited a number of times and was awarded PI's first Distinguished Research Chair.

Masterminds

In 1988's Masters of the Universe, *Stephen Hawking, Arthur C. Clarke, and Carl Sagan (not pictured) joined Magnus Magnusson, presenter of the BBC's* Mastermind *quiz show, to discuss life, the Universe, and everything.*

Geek heaven

In 2000, Hawking appeared as a guest star in the animated television series Futurama, *along with Nichelle Nichols (Lt Uhura from the original Star Trek),* Dungeons & Dragons *creator Gary Gygax—and, in the foreground, Al Gore.*

String fest

Hawking attends a press briefing at the 2006 International Conference on String Theory in Beijing, China. Next to him, right to left, are physicists Shing-Tung Yau, Edward Witten, David Gross, and Andrew Strominger.

Marilyn and me

Hawking pictured in his office at the Department of Applied Mathematics and Theoretical Physics (DAMTP) in 2001. Hawking once wrote that if time travel ever became possible he would drop in to visit Marilyn Monroe.

Royal honors

Hawking meets Britain's Queen Elizabeth II during the Chelsea Flower Show at Royal Hospital Chelsea, London, in 2010. He has been honored twice by the Queen, becoming a CBE in 1982 and a Companion of Honour in 1989.

Bestsellers

Since the publication of A Brief History of Time *in 1988, Hawking has enjoyed success as a bestselling popular-science author. He is pictured here promoting his book* The Universe in a Nutshell *in 2001.*

Stephen of the Antarctic

One of Hawking's most exceptional qualities is his refusal to let his disability hold him back. In 1997, he took this to a new level by visiting Antarctica.

Pioneer over G

In 2007, Hawking took a zero-G aircraft flight aboard a Boeing 727. The flight was made in preparation for his ultimate goal—to travel into space with Richard Branson's Virgin Galactic space tourism company.

Physics icon

Hawking next to a bust of himself unveiled at the Centre for Theoretical Cosmology in Cambridge on December 20, 2007. The bronze bust was commissioned to mark the opening of the center, set up to answer the Universe's biggest questions.

Genius at work

Stephen Hawking where he is most at home—in his Cambridge office at DAMTP, contemplating mathematical equations and the laws of the Universe. This photograph was taken in September, 2007.

Chapter 1

Life

In the Beginning

A stellar start

Stephen Hawking was born on January 8, 1942, which coincidentally marked the 300th anniversary of the death of **Galileo Galilei**, the Italian physicist, astronomer, and mathematician who is widely recognized as one of the founders of modern science. For his contentious views and scientific discoveries, Galileo was tried by the Roman Catholic Church and placed under house arrest in Florence, during which time he wrote some of his finest works. Was it remarkable that one of the world's **most eminent cosmologists** was born on the day that marked the anniversary of Galileo's death? With his signature dry wit, Hawking likes to point out that 200,000 other babies came into the world that day, too, so perhaps it wasn't such a coincidence after all.

Wartime upheaval

Stephen's parents, **Frank and Isobel Hawking**, lived in the leafy suburb of Highgate in north London. In 1942, the British capital was being bombarded by the Luftwaffe every night, with devastating loss of life and damage to homes and factories. So Frank and Isobel decided that **Oxford**, which had escaped the bombing, was a much safer place to bring their precious bundle into the world. Years later, Hawking would return to take up a scholarship at University College, Oxford, to read natural sciences. This marked the beginning of his journey to uncover some of the most fascinating secrets of the Universe.

Meet the parents

Frank and Isobel were from middle-class backgrounds and both attended **Oxford University**. Most Oxbridge undergraduates at the time fell into one of two categories: the privately educated alumni of schools such as Eton and Harrow whose wealthy families funded their champagne lifestyle; or the bright ranks of lesser-known independent schools and grammar schools, who wore shabby tweeds and couldn't afford to party. Frank and Isobel belonged in the latter category. After studying **medicine** at University College, Frank specialized in tropical diseases, and his research took him to East Africa before the outbreak of war. Frank wanted to enlist but his medical and research skills were considered of great importance in Britain, so he was persuaded to stay and work at a medical institute where he met Isobel and love blossomed. They married, and Stephen was soon on his way, later joined by **Mary, Philippa**, and an adopted brother, **Edward**.

3-Second Brief
Hawking's mother gives birth in Oxford to escape the bombs that were raining down on London during the Blitz. His birth coincides with the 300th anniversary of Galileo's death.

Related Thoughts
see also
CHILDHOOD
page 22
OXFORD DAYS
page 28

❝Galileo, perhaps more than any other single person, was responsible for the birth of modern science.**❞**

Galileo

Oxford

Frank Hawking with
son Stephen, 1942

Childhood

Schooldays

The Hawking family moved to a large rambling family home in **St. Albans, Hertfordshire**, when Hawking was eight years old. His father wanted him to attend one of the finest independent schools in the land and entered him for the highly competitive Westminster School in London. However, on the day of the entrance examination, Hawking was ill, and that was the end of that. Instead he attended St. Albans School, which was also considered **academically excellent** if not as prestigious as Westminster. It had close ties with the city's magnificent cathedral and educated 600 boys to Advanced (A) level. Hawking sailed through the entrance exam and joined at the **age of 11**. It was to prove the ideal academic environment for a budding genius such as he.

At home with the Hawkings

The Hawking family home at 14 Hillside Road was a **clutter of books**, paintings, and treasured objects that the family had collected on journeys to exotic destinations. Wallpaper peeled away in many places, there were holes in the walls, and the carpets weren't replaced until they were worn through. The Hawkings' family car was a London **black cab** which was replaced by a Ford Consul when Frank and Isobel decided to take a trip across Europe and Asia to India. The whole family went on the year-long adventure, apart from Stephen, who had to stay at home and pursue his studies. It was in this cultured environment that his **imagination** was allowed to take flight.

"A figure of fun"

Hawking was the archetypal "**schoolboy swot**," looking awkward, eccentric, and disheveled in his smart new school uniform. He was always a skinny child and showed little interest in the rough and tumble of the sports field. His slight lisp and tendency to jabber **isolated** him further, and he became a target of classroom teasing. His talents weren't always evident to friends and fellow pupils in his early years; when he was 12 years old, one of his friends bet a bag of candy that he would "never come to anything." Hawking observes with modesty: "I don't know if this bet was ever settled and, if so, which way it was decided."

3-Second Brief
Hawking grows up in a bookish environment and becomes the focus of schoolboy teasing for his image as a "swot."

Related Thoughts
see also
IN THE BEGINNING
page 20
SCIENTIFIC BETS
page 102

❝It doesn't matter what school you went to or to whom you are related. It matters what you do.**❞**

Stephen Hawking with sisters Mary and Philippa

St. Albans School

Early Promise

Bright young things

Hawking was in the top class of his year at St. Albans and was considered bright, if not **brilliant**, by his teachers. He became part of a clique of clever boys who shared his enthusiasm for schoolwork and cerebral pursuits. Instead of bopping and jiving to rock 'n' roll with other teenagers during the 1950s, Hawking and his friends preferred classical music and attended concerts at the **Royal Albert Hall** in London. They read the works of William Golding, Aldous Huxley, and Kingsley Amis, and revered the intellectual Bertrand Russell. Despite an avalanche of homework each night and compulsory sports at weekends, the boys found time to hold genteel sherry parties and take **long bicycle rides** in the Hertfordshire countryside.

Boffin's bedroom

Part messy teenager's retreat, part mad professor's laboratory, Hawking's room at the top of the house was a clutter of model aircraft, piles of unfinished homework, and schoolbooks. Signs of his **growing passion for science** lay all around—from the curious gadgets made out of metal and wire to the racks of test-tubes which held discolored liquids formed from **long-abandoned experiments**. It was here that Hawking and his friends would meet to indulge their passion for inventing board games. Hawking, the logician, would devise the rules of each game, but often they were so complicated that even his brainy chums struggled to decipher them.

An early computer

Hawking sailed through his Ordinary (O) level examinations, gaining 10 in total, and selected **mathematics, physics**, and **chemistry** to study at A level. In his final two years at school, he and his friends **built a computer** called LUCE, the Logical Uniselector Computing Engine. At the time in Britain, only a few universities and the Ministry of Defence had computers, so it was quite a feat. The boys used old parts from an office telephone exchange and created a complex labyrinth of electrical connections. **LUCE** was primitive, but it worked, and the only major problem its inventors encountered was bad soldering. Long after the boys left, a new Head of Computing found a box under a desk labeled LUCE. He thought it just a mass of useless junk and only years later realized what a blunder he had made, binning the computer built by Stephen Hawking.

3-Second Brief
As a teenager, Hawking becomes part of a clique of brainy boys who devise complex board games and build a primitive computer.

Related Thoughts
see also
WHY PHYSICS?
page 26
POPULAR SCIENCE CHAMPION
page 124

❝My classwork was untidy and my handwriting was the despair of my teachers. But classmates gave me the nickname Einstein, so presumably they saw signs of something better.❞

Stephen Hawking, aged 12

Royal Albert Hall,
London

Why Physics?

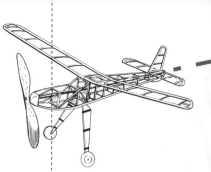

Boyhood inventions

In his teens, Hawking loved to build **model airplanes** and boats, despite the fact that he wasn't particularly good with his hands. Tellingly, he has said: "It was always my aim to build working models that I could control. I didn't care what they looked like. I think it was the same drive that led me to invent a series of very complicated games with friends. These games, as well as the trains, boats, and airplanes, came from the urge to know how things worked and to control them." Since Hawking began his **PhD**, this desire has been met by his research into cosmology. As he says: "If you understand **how the Universe operates**, you control it in a way."

The desire to fathom space

Stephen's father, Frank, encouraged his son's interest in science and wanted him to follow in his footsteps to study medicine. In his early years, Hawking didn't differentiate between one science or another, but from his teens he knew that he wanted to study physics because it was the most "fundamental science." The irony is, he found **physics** so "easy and obvious" that he said that it was the most "boring" subject at school. For him, chemistry was much more fun because unexpected things, such as **explosions**, kept happening. Despite finding physics so easy, he felt that the study of its laws, allied with astronomy, offered the hope of understanding where we came from and **why we are here**. Finding the best way to address such questions in his research would not always be so clear.

The inspiring Mr. Tahta

On becoming famous, Hawking was asked to name one teacher who had inspired him. His response was "Mr. Tahta." He was referring to **Dikran Tahta**, his mathematics teacher at St. Albans School in the 1950s, who rose to become a highly respected figure in the world of mathematics. An outstanding and somewhat maverick teacher, Tahta fostered a lively, interactive atmosphere in his classes, and his passion for his subject was infectious. On becoming a Lecturer in Mathematics at **Exeter University**, Tahta would encourage postgraduates to make animated films of the nearby countryside and bake cakes as part of their study. He wanted to free the "mathematics psyche" from narrow, restrictive modes of thought—an approach that would become vital for Hawking's future study of cosmology.

3-Second Brief
Hawking's desire to understand how toys and games work, and are controlled, starts a lifelong fascination with physics.

Related Thoughts
see also
POSTGRADUATE LIFE AT CAMBRIDGE
page 30
COSMOLOGY
page 84

"I wanted to fathom the far depths of the Universe."

Oxford Days

Going up to university

In 1958, tensions rose in the Hawking household over selecting a university course for Stephen. Frank wanted him to follow in his footsteps by reading medicine at **University College, Oxford**, but Hawking wanted to study natural sciences—a course combining mathematics and physics. Eventually he agreed to apply to University College—but for the course of his choice. The entrance process was grueling, with two days of examinations and an interview with senior members of the college. An anxious wait ensued, but there was no need for chewed fingernails. He had scored around **95 percent** in the physics examination, and the college offered him a scholarship. The only condition was that he achieve two A level passes—which was something of a formality.

Dreaming spires

Hawking "went up" in 1959, at the age of 17. His early days at Oxford were somewhat aimless, as he found the physics coursework so easy. In the first year he only attended mathematics lectures and seminars with his tutor, **Dr. Robert Berman**. Fellow students would remark that it took them a week to tackle the physics problems that Hawking could solve in minutes. It would have been easy for him to slide into a life of apathy. Discovering an interest in the popular university sport of **rowing** gave him fresh focus in his second year. Being so light, Stephen was perfect as a coxswain, the person who steers the boat and regulates the stroke rate. Crews worked hard for long hours on the river. They played hard, too, enjoying **alcohol-fueled parties**, in which Hawking took an enthusiastic part.

The final countdown

In his third year at Oxford, Hawking applied to study for a PhD in cosmology at **Cambridge University** and needed to pass his final examinations with first-class honors. Berman knew that despite Hawking's natural talent, lack of work could cost him dear. He was right. Hawking was **borderline** between the First that he needed for Cambridge and a Second. The examiners decided to give him a final "viva voce" (oral) exam to **decide his fate**. As Berman has since said: "They had the intelligence to realize that they were talking to someone far cleverer than most of themselves." A dry comment from Hawking could also have swung their decision. He said: "If you award me a First, I will go to Cambridge. If I receive a Second, I shall stay in Oxford, so I expect you will give me a First." They did.

3-Second Brief
Finding the coursework at Oxford "too easy," Hawking becomes lazy and nearly fails to get the qualification he needs for postgraduate study at Cambridge.

Related Thoughts
see also
POSTGRADUATE LIFE AT CAMBRIDGE
page 30

A GATHERING STORM
page 32

❝The prevailing attitude at Oxford was very anti-work. You were supposed to be brilliant without effort or to accept your limitations and get a fourth-class degree.❞

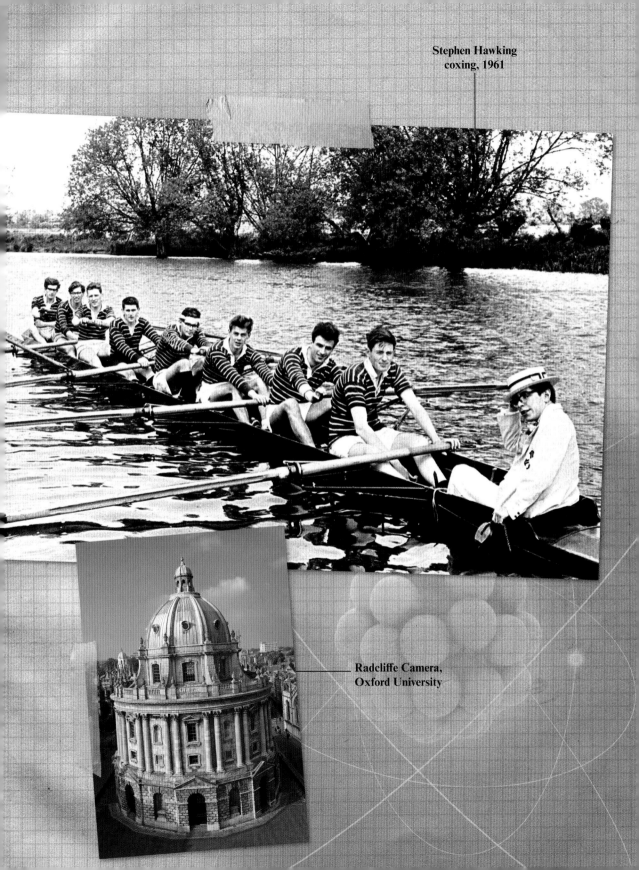

Stephen Hawking
coxing, 1961

Radcliffe Camera,
Oxford University

Postgraduate Life at Cambridge

Why cosmology?

At the time of applying for his PhD, there were two options open to Hawking: elementary particles—the study of the very small—and cosmology, which lay at the opposite end of the scale. He decided on cosmology because there was a clearly defined theory to follow—the **general theory of relativity**, which Einstein had established—and yet there was so much still to discover. In 1962, when Hawking graduated, Oxford University did not offer cosmological research, so he applied to Trinity Hall, at Cambridge University, where he believed that he would study under **Fred Hoyle**, the most eminent scientist in the field at the time. Hawking was in for a big disappointment—at least at first.

Guiding light

Instead of getting Hoyle as his PhD supervisor, Hawking was assigned to **Dennis Sciama**. At first Hawking thought this was a disaster—he had never heard of Sciama—but soon he realized what a blessing it was. Fred Hoyle spent a lot of time overseas and could only give limited attention to PhD students. Sciama was a **fine cosmologist** who was in the process of fostering some of the greatest astrophysicists of the age, and he had plenty of time for the young Hawking—who was struggling with his studies again. His laziness as a student at Oxford left him poorly prepared to understand the complex mathematical equations associated with relativity. Sciama knew that Hawking was an **exceptionally gifted** student and that he could flourish—he just needed to find the right subject.

The eureka moment

That subject came along in 1964, after Sciama and his cosmology students attended a talk given in London by gifted mathematician **Roger Penrose**. Hawking, like the others, was intrigued by Penrose's idea that a spacetime singularity (a point of infinite density) could exist at the center of a **black hole**. After mulling over the concept on the train home to Cambridge, Hawking turned to Sciama and said: "I wonder what would happen if you applied Roger's singularity theory to the entire Universe." This was the turning point in Hawking's career. It made the final chapter of his PhD a brilliant piece of work and opened up a **revolutionary** new field of cosmology research.

3-Second Brief
Hawking's laziness at Oxford is exposed at Cambridge where he struggles with his PhD research.

Related Thoughts
see also
CAMBRIDGE, A LOVE OF HAWKING'S LIFE?
page 56
SINGULARITY THEOREMS
page 74
COSMOLOGY
page 84

❝I found that elementary particles were less attractive because, although they were finding lots of new particles, there was no proper theory. In cosmology, there was ... Einstein's general theory of relativity.**❞**

Cambridge, England

Roger Penrose

A Gathering Storm

Early warning signs

During his final year at Oxford University, Hawking became **clumsy** and fell a couple of times for no reason. At Cambridge, he began to find it difficult to tie his shoelaces, and he developed a slight speech impediment. He returned to his parents' home in St. Albans for the Christmas holiday, and Frank was quick to notice the **physical changes** in his son. Hawking was referred to a specialist and in spring **1963** underwent tests in hospital. This involved taking a muscle sample from his arm, placing electrodes into him, and injecting radio-opaque fluid into his spine and watching its movements using X-rays. The medics said he was an atypical case and advised him to return to Cambridge and take vitamins until they could identify what was wrong.

The dreadful news

The diagnosis soon came, and it was the worst possible news. He had developed **Amyotrophic Lateral Sclerosis** (ALS), a rare and incurable disease known more commonly in the USA as **Lou Gehrig's** disease and in Britain as motor neurone disease. ALS leads to a gradual disintegration of the nerve cells in the spinal cord and paralysis as the muscles atrophy. The initial signs are weakness, slurred speech, and difficulty in swallowing. The **brain is unaffected**, however, leaving thought, memory, and feeling untouched. Death usually comes as a result of suffocation or pneumonia and, although the symptoms are painless in the final stages of the disease, patients are often prescribed morphine in order to alleviate feelings of fear and depression.

Facing the future

Hawking was given a life expectancy of two years. He sank into a **deep depression** and shut himself away from life at Cambridge. Speculation that he indulged in heavy drinking binges at this time have been refuted by Hawking, who has remarked: "Reports in magazine articles that I drank heavily are an exaggeration. I felt somewhat of a tragic character. I took to listening to Wagner." An experience in hospital while he was undergoing further tests **helped him find the courage** to face his illness—he saw a boy die of leukemia in the bed opposite. Hawking has since said that whenever he felt sorry for himself, he remembered that boy. A developing romance and exciting new focus in his work were also to spur him on.

3-Second Brief
Hawking is given the dreadful news that he has Amyotrophic Lateral Sclerosis (ALS) and is told he has just two years to live.

Related Thoughts
see also
LIVING WITH ALS
page 38
HOW HAWKING WORKS
page 44

❝Although there was a cloud hanging over my future, I found to my surprise that I was enjoying life in the present more than I had before. I began to make progress with my research.**❞**

Stephen Hawking at
his Oxford University
graduation

Falling in Love

Magnetic attraction

Jane Wilde was just a teenager studying for A levels in St. Albans when she met the slightly eccentric Cambridge postgraduate Stephen Hawking at a **New Year's Eve party in 1962**. She was drawn to him and although she remembers sensing an intellectual arrogance about Hawking there was also "something lost" about him, something happening that "he wasn't in control of." At this time, he was experiencing increasing physical difficulties and was about to undergo specialist tests in hospital. Jane had a place to study modern languages at Westfield College, University of London, the following year and was busy with A levels. After the diagnosis, however, the pair began to see more of each other and a **deep relationship developed**.

Romantic Cambridge

After going up to college, Jane would spend **idyllic weekends in Cambridge** with Stephen and his friends, enjoying picnics by the river and watching other students gliding by on the city's famous **gondola-like punts**. They became engaged, and this marked a huge turning point in Hawking's life. His research began to go from strength to strength and his natural sense of buoyancy returned. He has remarked several times that getting engaged to Jane changed his life and **gave him something to live for** after being diagnosed with ALS. Hawking knew that in order to marry Jane he would have to get a job or a university fellowship. Without this, it would be difficult to provide stability for them or to continue his evolving research into black holes.

Jane becomes his rock

Hawking applied for a fellowship at **Gonville and Caius** (pronounced "Keys") College, Cambridge. By that time, his condition meant that he could no longer write, so he arranged for Jane to come to Cambridge and type out his application. He met her off the train and, on seeing her arm in plaster due to an accident the previous week, found it hard to conceal his disappointment. She was able to write the application out longhand, however, and a friend typed it for them. The couple enjoyed a wonderful year in **1965**. Hawking, who had by that time received his doctorate, became a fellow at Gonville and Caius, and he and Jane were married at **Trinity Hall**. Jane knew that her husband might have only a few years to live, but the couple faced the future with a positive attitude and decided to start a family as soon as possible.

3-Second Brief
Love blossoms when Hawking meets Jane Wilde at a New Year's Eve party. Jane becomes his rock and helps him come to terms with having ALS.

Related Thoughts
see also
CHILDREN
page 36
LIVING WITH ALS
page 38

❝Without the help that Jane has given I would not have been able to carry on, nor have the will to do so.❞

Stephen and his first wife Jane

—— **Romantic Cambridge**

Children

Against all odds

In 1967, Stephen and Jane welcomed their first son, **Robert**, into the world—four years after doctors had told Hawking that he had just two years to live. Robert's birth was another pivotal point in Hawking's life. He was **working at a phenomenal rate**, his reputation as a brilliant physicist was growing, and every paper that he published seemed to break new barriers. Becoming a father was the icing on the cake, and, as Jane observed at the time, being responsible for this "tiny creature" gave him added impetus in life. That tiny creature was to **follow in his famous father's footsteps** by studying physics at Cambridge and at postgraduate level before moving to the USA.

Home life

In 1970, Robert was joined by **Lucy**, and the Hawkings were clearly delighted with their expanding family. This put added strain on Jane, however, who had to juggle the demands of bringing up young children with running a home and caring for her husband. When their third child, **Timothy**, was born in 1979, life at home in Cambridge became a hectic whirl of **work, children**, and **socializing**. It was Jane who would teach Robert, Lucy, and Timothy how to play games such as cricket and croquet. Hawking would play tag with them, racing around the lawn in the wheelchair to which he was by that time confined, showing impressive dexterity. One of his great regrets, however, is that he wasn't able to play more games with his children as they were growing up.

Cosmic adventures

The Hawkings educated their children at independent schools in Cambridge, and the increasing cost of this, as well as nursing care, was part of the impetus behind Hawking's idea of writing a popular-science book. Stephen's daughter Lucy studied modern languages at Oxford and carved a successful career as a newspaper and radio journalist. Together father and daughter have co-authored a series of three action-packed **children's books** about a young cosmic adventurer called George. The tales marry exciting plots with intriguing facts about astrophysics in order to stimulate curiosity in the minds of young readers. The books are hugely popular and have been translated into **37 languages**. In 2010, Lucy Hawking took up a year's residency as a writer at **Arizona State University**.

3-Second Brief
Four years after the dreadful diagnosis, the Hawkings become parents to Robert, who is followed by Lucy and Timothy. Hawking perfects the art of playing tag in a wheelchair…

Related Thoughts
see also
A GATHERING STORM
page 32

FALLING IN LOVE
page 34

"I have had motor neurone disease for practically all my adult life. Yet it has not prevented me from having a very attractive family.**"**

Stephen Hawking and family, Cambridge, 1981

Living with ALS

Battle with bureaucracy

At the time of his marriage Hawking could still walk, albeit with the aid of sticks. It was vital for the couple to find a house centrally located in Cambridge, so they applied to Gonville and Caius College to see if it could offer some help. The Bursar rather sniffily informed them that it was not college policy to help fellows with housing. Instead, they had to wait and search, sometimes sleeping in a student hostel, until a house became available to rent in **Little St. Mary's Lane**, just minutes from the center of Cambridge. The house was ancient and tiny with a narrow, winding staircase. With **characteristic stubbornness**, he would insist on getting upstairs on his own, even though it took much time and effort.

ALS tightens its grip

The severity of Hawking's condition remained fairly constant from diagnosis until 1965 when, on a trip to **Miami** to give a talk on relativity, his speech suddenly deteriorated to a slur and a colleague had to step in to help. By the late 1960s, he was forced to use a **wheelchair**, a situation that he had fought for years. He refused to let this limit him and became a familiar sight around Cambridge, speeding along the pavements on his way to deliver a talk, have lunch at Gonville and Caius, or meet with students he was supervising. Showing his more irascible side, Hawking will sometimes use his wheelchair in order to get a point across. If someone is annoying him, or tactlessly bemoaning their own ailments, he has been known to run over their toes.

Student in residence

By the early 1970s, Hawking was having difficulty getting in and out of bed and feeding himself. Although in **1974** they moved into a ground-floor apartment in West Road, Cambridge, it was clear that Jane, with two young children to take care of, would need more support. The couple decided to offer free accommodation to a student in return for their help. The role of "**graduate assistant**" was to become coveted among Hawking's research students, and **Bernard Carr**, one of the earliest to take on this role, said "it was like participating in history." Students mucked in with family life, babysitting the children and helping to organize trips and schedules. In return, they gained insight into the mind of a genius, and for some it became a great career move.

3-Second Brief
The newly married Hawkings gain no concessions over housing from the college, despite Stephen's condition. As the 1960s progress, he finally accepts that he must use a wheelchair.

Related Thoughts
see also
FALLING IN LOVE
page 34
CALIFORNIA
page 42

"I was fortunate that my scientific reputation increased at the same time that my disability got worse. People were prepared to offer me a sequence of positions in which I only had to do research without having to lecture."

Little St. Mary's Lane, Cambridge

The Next Einstein?

Rising fame

By the late 1960s and early 1970s, an air of mystique began to grow around Hawking at Cambridge, and quadrangles echoed with whispered speculation about the gauche genius and his work. Hawking and mathematician **Roger Penrose** progressed further on their theories regarding black holes, and physicists across the world began to sit up and take note. Their work led to fresh insight into how the outer surfaces, or "horizons," of these mysterious objects behave, gave birth to the new field of **black hole thermodynamics** and led directly to the groundbreaking theory in 1974 that black holes could emit radiation. **Awards and accolades** began to flood in for both men.

The Royal Society

Hawking's findings at the time challenged all conventional thought in cosmology. In 1974, his research into black holes and quantum gravity earned him such respect that he was awarded one of science's greatest honors: he was invited to become a Fellow of Britain's prestigious **Royal Society**. This is a title bestowed only to the world's most eminent scientists. Past recipients of the "FRS" have included Isaac Newton, **Charles Darwin, Albert Einstein, Ernest Rutherford**, and more than 80 Nobel Laureates. Joining at the age of 32, Hawking became one of its youngest ever members.

Lucasian Professor of Mathematics

In 1979, Cambridge University gave Hawking one of its highest accolades by appointing him **Lucasian Professor of Mathematics**, widely regarded as one of the world's most prestigious chairs. It was deeded to Cambridge University by English politician Henry Lucas in 1663, and the list of those who have occupied the position includes some of the fathers of modern science. Physicists as well as mathematicians can be appointed to the chair, and its incumbents have included **Isaac Newton, Charles Babbage**, and **Paul Dirac**. Hawking was only 37 years old when he became Lucasian Professor and already some contemporaries were hailing him as "**the next Einstein**." They believed he was the scientist most likely to find a "theory of everything" by resolving the disparities between general relativity and quantum mechanics to present a unified description of the forces of nature.

3-Second Brief
Hawking's fame grows, if not his fortune, and he becomes one of the youngest Fellows of the Royal Society.

Related Thoughts
see also
BLACK HOLES
page 70
KNOWING THE MIND OF GOD
page 94
CONTEMPORARIES
page 116

❝My goal is simple. It is a complete understanding of the Universe, why it is as it is, and why it exists at all.❞

California

A scientist's dream

By 1974, Hawking's reputation as a brilliant cosmologist was growing worldwide, and he was invited to spend a research year funded by a Sherman Fairchild Distinguished Scholarship at **California Institute of Technology** (Caltech). Based in Los Angeles, Caltech may be small, educating just 2,150 students a year, but it is a center of excellence, and 31 of its alumni and faculties have won the **Nobel Prize**. Its facilities are the stuff of scientists' dreams, with access to some of the finest telescopes in the world, the massive Jet Propulsion Laboratory annexed to the campus, and a never-ending supply of funds pouring in from sponsors. Hawking particularly relished the chance to conduct research alongside the eminent theoretical physicist **Kip Thorne**.

The LA lifestyle

The Hawking family were excited at the thought of a year in sunlit California. Ever-sociable, they loved to have guests to stay and would take them on trips along the **Pacific Highway** or to the beautiful desert around **Palm Springs**. The family loved the sunny LA lifestyle which must have come as a welcome change from the long winter months of Cambridge. Sheltered by campus life, the Hawkings had the support that they needed, and **Caltech**, unlike Cambridge at times, treated Hawking and his needs with thoughtfulness and respect. Wooden ramps were fitted against the curbs near his department, and he was given a new office with all the aids and materials that he required. His condition was worsening, and he needed increased nursing care from this time onward.

Great minds

With the addition of Hawking—the "bright young thing" from Cambridge—Caltech boasted an enviable lineup of physicists. Kip Thorne was head of the relativity group, Nobel prize-winner **Richard Feynman** was working there at the time, and they were surrounded by the **crème de la crème** of American physics students. Thorne, like Hawking, has the ability to communicate both the excitement and significance of discoveries in cosmology to scientific and lay audiences alike. Through him, Hawking met a physicist who would become a lifelong friend, **Don Page**. The two instantly hit it off and worked on a key paper on black holes during Hawking's research year. Hawking clearly loved his time at Caltech because he has returned many times to his "home from home" there, often receiving a superstar reception.

3-Second Brief
The Hawking family exchange the grey skies of Britain for the blue of California when Stephen takes up a prestigious position at Caltech—a center of scientific excellence.

Related Thoughts
see also
CONTEMPORARIES
page 116

"It was only when we went to California in 1974 that we had to get outside help, first a student living with us, and later nurses."

Kip Thorne

Caltech, Los Angeles

How Hawking Works

Mind over matter

Hawking has been hailed as the most brilliant theoretical physicist since Einstein. So, just how intelligent is he? Hawking is playfully dismissive of the traditional **Intelligence Quotient** (IQ) test, quipping in an interview that "people who boast about their IQ are losers." However, he did go on to add that he hoped that his "would be high." Many people have commented on Hawking's **phenomenal memory**, which is one of the facets that has allowed him to overcome his disability. As a student, he rarely needed to keep notes, and a secretary who worked for him was amazed not only by his ability to dictate 40 pages of equations from memory but by his recall the next day of a mistake he had made. Add to this Hawking's gift for theoretical physics and you have something very special.

Collaboration is key

If he is asked about his mode of working, Hawking often says that he relies heavily on **intuition**, thinking that an idea "ought to be right." He will then try to prove it—sometimes he is wrong, sometimes the original idea is flawed, but that leads on to new ideas. Hawking enjoys collaborating and discussing his work with other fellows and students at Cambridge's Department of Applied Mathematics and Theoretical Physics (**DAMTP**). The daily routine at the department has continued for years. Hawking and his PhD students always break for morning coffee and afternoon tea to discuss ideas and have a joke. Writer **Dennis Overbye** visited Hawking there and described the students gathered around him as resembling "in age, dress, pallor, and evidence of nutritional deficiency the road-crew of a rock and roll band."

The command center

As Hawking's ALS has progressed, his wheelchair and office at DAMTP have been adapted accordingly. His wheelchair is **powered by batteries** fitted under the seat, which also supply the computer attached to his chair. The whole computer is mounted to the arm of the chair so that Hawking can see the screen easily. To speak to people on the telephone he uses Voice over Internet Protocol (VoIP) software or connects his chair computer directly to a phone socket. He can then issue digital commands from his computer instructing the phone system to dial a number, answer the phone, or hang up at the end of a call. A programmable **infra-red remote control** attached to his computer enables him to open doors and operate lights, televisions, and other appliances.

3-Second Brief
Hawking's astonishing memory allows him to visualize and manipulate complex systems of mathematical equations in his head, without writing a single formula down.

Related Thoughts
see also
DISABILITY TECHNOLOGY
page 122

"I find it a great help to discuss my ideas with other people. Even if they don't contribute anything, just having to explain it to someone else helps me sort it out for myself.**"**

Hawking's Beliefs

Cosmology and religion

"Heaven… is a fairy story for people afraid of the dark." So says Hawking in a comment that sums up his attitude toward religion. Like Einstein, he finds faith and religion to be at odds with his view of the Universe. He will concede that it's difficult to discuss the origin of the Universe without mentioning God, but for him religious beliefs remain a personal choice and **mathematical reasoning** overrides any need for spirituality. His **agnosticism** has led to hurt, tension, and frustration for his ex-wife Jane, who is a Christian believer. She has said that, without her faith in God, it would have been difficult for her to marry Stephen in the first place and find the optimism to keep going.

Hawking the activist

Hawking was brought up in a politically active household. His mother Isobel was a member of the **Communist Party** in the 1930s before joining **Labour** (Britain's major left-wing political party), and she encouraged him to go on demonstrations with her, including those supporting nuclear disarmament. Hawking has actively supported Labour but was openly critical of policies that involved cuts to science funding. As his fame has increased, he has become more vocal about social and political issues, speaking out on matters as diverse as the senselessness of nuclear weapons, the **plight of the poor**, and on the environment. He also led a campaign to overthrow a ruling preventing women from entering Gonville and Caius College at Cambridge.

Campaigner for disabled rights

Hawking tends to play down his disability, wishing to be "defined" by his achievements and personality rather than the physical challenges of his life. He is a keen campaigner for the rights of other disabled people, however. In the 1970s, he had a **long-running battle** with Cambridge University over who should pay for disabled access to DAMTP. Hawking eventually won, and this seemed to fire him up, because ever since he has campaigned actively for improved access for disabled people, in Cambridge in particular. In 1979, the **Royal Association for Disability and Rehabilitation** nominated Hawking as "Man of the Year."

3-Second Brief
Hawking is open about his views on religion and politics, even if it means courting controversy.

Related Thoughts
see also
RIGHTS FOR THE DISABLED
page 120
RUFFLING FEATHERS
page 128

❝We are such insignificant creatures on a minor planet of a very average star in the outer suburbs of one of a hundred thousand million galaxies. So it is difficult to believe in a God that would care about us or even notice our existence.**❞**

A New Voice

A brush with death

In **1985**, Hawking came very close to losing his battle with ALS. He was conducting research at **CERN** (the European Organization for Nuclear Research) near Geneva, Switzerland; and A Brief History of Time had not yet been published. Jane was staying with friends in Germany at that time, and Stephen was cared for full-time by nurses. One night in August, his nurse found him struggling to breathe, and he was rushed to the Cantonal Hospital to be put on a ventilator. Doctors suspected a blockage in the windpipe and pneumonia, which is often fatal for people with ALS. They recommended a **tracheotomy** which would involve cutting a hole in his neck (a tracheostomy) and implanting a breathing device. The operation was essential but it meant that Stephen would never "speak" again.

Dark times

The decision on whether to go ahead rested with Jane. It was a **huge dilemma**. On the one hand, his speech had already become almost unintelligible to all apart from his family and close friends. On the other, what kind of life could he lead without the ability to speak at all? Agreeing to the operation would also bring additional financial pressures. He would need round-the-clock nursing care, and **Britain's National Health Service** could only provide care for a few hours a week. Also, how would he work when his only mode of communication would be to blink his eyes to choose words written on a card held in front of him? As Jane gave her consent, the future for the Hawking family looked bleak.

New hope

After Hawking had returned to Cambridge to recuperate, he was given a significant boost when American computer expert **Walt Woltosz** sent him a program he had written called **Equalizer**. This enabled Hawking to choose words on a computer screen by squeezing a switch held in his hand. (Later he would twitch his right cheek, activating an infrared beam from a device mounted on his spectacles to move a cursor through his dictionary.) When he had created a sentence it would be sent to a **voice-synthesizer** that "spoke" it for him. Hawking found the program slow (it could only process around 10 words per minute), but he no longer needed an interpreter and a portable version of the voice synthesizer could be attached to his wheelchair. The Americanized, slightly Scandinavian, voice would soon become recognized the world over.

3-Second Brief
Hawking becomes dangerously ill in Switzerland and requires a tracheostomy to survive, which robs him of speech. Computer wizardry allows him to speak again with the now-familiar "Hawking" voice.

Related Thoughts
see also
LIVING WITH ALS
page 38

❝Mankind's greatest achievements have come about by talking, and its greatest failures by not talking. It doesn't have to be like this.❞

Stephen Hawking at Cambridge

CERN, Geneva, Switzerland

A Brief History of Time

Financial pressures

By the mid-**1980s**, Stephen and Jane faced the mounting costs of nursing care and private school fees. There was always a fear lurking in the back of their minds: what if his condition made it impossible for him to continue work and he was forced to go into a home? Hawking had a plan, but it would take years to come to fruition. Prior to 1984, his papers and books had been well received in academic circles, but they were so complex that even many theoretical physicists struggled to decipher their meaning. Hawking was planning a **popular-science** work that would appeal to people from all walks of life. Cambridge University Press was interested and offered an advance of £10,000 ($18,000).

Across the pond

Some 5,000 miles (8,000 km) away, a New York literary agent, **Al Zuckerman**, had read an article about Professor Hawking and was intrigued by him. He contacted Hawking to begin discussions over a possible book, and Hawking enthusiastically replied, sending him a draft proposal and sample section of what was to become *A Brief History of Time*. The proposal generated such a buzz among publishers that Zuckerman was able to auction the book and secure a **$250,000** advance from Bantam Books for publication in the USA and Canada. Hawking began writing in 1984 and collaborated closely with editor **Peter Guzzardi**, who helped translate Hawking's complex vision of the cosmos into language that was understandable by the lay person. An introduction by the renowned astrophysicist Carl Sagan added to the book's appeal.

A runaway success

A Brief History of Time: From the Big Bang to Black Holes was not ready for publication until spring 1988, and Hawking nearly died of pneumonia before the manuscript was finished. The completed book was a challenging read, but it became an **instant success** when it was published in the USA and the UK, where it remained on the bestseller list for more than four years. Hawking became famous all over the world, and his book set the precedent for a generation of popular-science titles to come. To date, the book has sold **10 million copies worldwide**, although there are some who would question how many readers have actually read it from cover to cover.

3-Second Brief
Necessity is the mother of invention. Faced with school fees and rising nursing costs, Hawking devotes years of attention to writing a popular-science book. The rest, as they say, is history.

Related Thoughts
see also
SCIENTIST AND SUPERSTAR
page 52
POPULAR SCIENCE CHAMPION
Page 124

❝I am pleased a book on science competes with the memoirs of pop stars. Maybe there is hope for the human race.❞

Scientist and Superstar

The fame game

A Brief History of Time brought **superstardom** for Hawking, and while he has embraced some aspects of his celebrity, he has firmly rejected others. For many years, the scientific community was baffled that Stephen had not become Sir Stephen, but, in 2008, he revealed that he had turned down such an honor more than a decade earlier. Perhaps the greatest legacy of his fame will be that he established a groundbreaking genre of popular-science writing and inspired an interest in cosmology among readers across the globe. Instead of shying away from the glare of publicity, Hawking has **embraced popular culture**, and in the 1990s he became a regular on television and in a range of newspapers.

Hawking as castaway

In 1992, Hawking joined a long list of eminent writers, scientists, performers, and politicians asked to appear on the much-loved **BBC Radio 4** program, *Desert Island Discs*. Guests are asked to choose the eight pieces of music they would take with them if they were cast away on a desert island. Showing his great **love of classical music**, Hawking's selection included pieces by Wagner, Beethoven, Puccini, and Poulenc, and his favorite choice was Mozart's "Requiem in D Minor." **The Beatles** song "Please Please Me" made his final list as did Edith Piaf's rendition of "Non, je ne regrette rien." Guests can choose one book to take with them, and Hawking opted for George Eliot's *Middlemarch*. His luxury item was crème brûlée.

Keep talking

Hawking's artificially generated voice became iconic following a number of appearances in the mid-1990s. It featured first in the **Pink Floyd** song "Keep Talking," which was part of the band's 1994 album *The Division Bell*, and on a British Telecom advertisement that delivered the same message. Hawking is also not averse to self-parody and made a debut guest appearance in **The Simpsons** in a 1992 episode entitled, "They Saved Lisa's Brain." Viewers loved it, and he has featured in a number of episodes since then. This may have backfired on him, however, for in a recent debate in the *Guardian* newspaper in Britain, physicist Brian Cox asked him what are the most common misconceptions about his work. Hawking replied that "people think I'm a Simpsons character."

3-Second Brief
Hawking becomes a worldwide celebrity, with memorable appearances on a Pink Floyd album, *Desert Island Discs*, and in *The Simpsons*.

Related Thoughts
see also
FILM AND TELEVISION
page 126
NAMED IN HONOR
page 148

❝Your theory of a donut-shaped Universe intrigues me, Homer. I may have to steal it.**❞**

HAWKING CAMEOS ON AN EPISODE OF *THE SIMPSONS*.

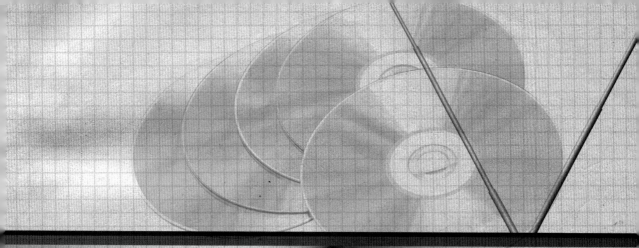

Globetrotting Genius

Personal pressures

Hawking achieved worldwide fame in the late **1980s**, and the blaze of publicity sparked by *A Brief History of Time* continued to burn unabated as he participated in interviews and documentaries. He maintained a **hectic schedule**, traveling across the world to deliver lectures and took a team of nurses with him. At home, marital pressures were building as Jane, who had supported and cared for Hawking for decades, felt more than ever that she lived under the shadow of his success. Jane had dressed, bathed, and fed him every day, while also bringing up a young family and completing a PhD. The **strain** was beginning to take its toll.

Second time around

David Mason, the computer engineer who had adapted Hawking's computer so that it could be fitted onto his chair, was married to a nurse called Elaine who helped to care for him during the late 1980s. She accompanied Hawking on a number of overseas trips, and the pair became close. In 1990, many in Cambridge were shocked when newspaper reports revealed that, after 25 years of marriage, Hawking had left the family home to move in with **Elaine**. The Hawking children found their parents' separation difficult to deal with, particularly Timothy, who was 11 at the time. Stephen divorced Jane in 1991 and married Elaine in 1995.

Overseas tours

During the 1990s, Stephen continued to juggle research, overseas tours, and responsibilities to students at DAMTP in Cambridge. His travels included a trip to Los Angeles where he met **Stephen Spielberg** who has followed Hawking's work with interest for a number of years. This trip was just one of a large number Hawking has found the energy to carry out and over the years he has traveled to India, Israel, South Africa, Canada, and even Antarctica, to name but a few. Elaine accompanied Stephen on many of his **overseas tours**, and the pair spent 17 years together, however in 2007 Hawking announced that the **marriage had come to an end** and he was starting divorce proceedings. He has not remarried since and remains in contact with Jane, his children, and grandchildren.

3-Second Brief
Hawking travels all over the globe on a series of lecture tours. His first marriage breaks down, when he leaves Jane for his nurse Elaine Mason.

Related Thoughts
see also
SCIENTIST AND SUPERSTAR
page 52

❝If I had a time machine... I'd travel to the end of our universe to find out how our cosmic story ends.❞

Cambridge, a Love of Hawking's Life?

A special place

Scientific institutes and universities across the world would pay a high price to recruit Hawking onto their permanent staff and bask in the reflected glory that his presence would bring. However, despite repeated stays at centers of excellence overseas, Hawking has remained faithful to Cambridge University and DAMTP. The city of Cambridge, with its **exquisite architecture** and unique beauty, has an atmosphere that has inspired and nurtured poets, scientists, and philosophers since medieval times. For Hawking, it's home for a number of deeper reasons: he courted Jane along the leafy "Backs" by the **River Cam**; they married in Cambridge; and their three children were born there. Hawking has fallen out with the university at times, particularly over disabled access, but the city remains his home and has a special place in his heart.

Bright social whirl

Hawking loves to socialize and has long been at the center of gatherings of **Cambridge intelligentsia**. The tiny home he owned with Jane in Little St. Mary's Lane would often be crammed with friends listening to classical music, preparing food, and "talking shop." Around the table sat a number of people who also studied under **Dennis Sciama** at the vanguard of cosmological research, and some would go on to greatness. **Martin Rees**, for example, became Britain's Astronomer Royal, Master of Trinity College, Cambridge, and President of the Royal Society—the most eminent group of scientists in Britain. Rees and Hawking have remained friends over the years, but they disagree over the debate between religion and science.

Cambridge honors one of its finest "sons"

In 1989, Cambridge University bestowed one of its most rare and prestigious honors on Hawking when he was made an honorary **Doctor of Science**. The award was presented by the University's Chancellor, HRH Prince Philip, and hundreds of people lined the route applauding as Hawking wheeled along **King's Parade** to the ceremony at Senate House. Hawking is regularly seen in restaurants, lecture halls, and concerts in Cambridge and often acknowledges a smile or wave from an admirer. He values his privacy, however, and when the world media wanted to dissect his marital problems and in particular his split from Elaine, Cambridge University, friends, and colleagues closed ranks.

3-Second Brief
Hawking and Cambridge go back a long way—through good times and bad—but the city continues to hold dear one of its finest sons.

Related Thoughts
see also
FALLING IN LOVE
page 34
CHILDREN
page 36

❝I came to Cambridge 47 years ago as a graduate student. Cosmology was a relatively unexplored field, but in the Department of Applied Mathematics and Theoretical Physics I found an atmosphere of adventure and discovery.❞

Stephen Hawking, Martin
Rees, and Michael Griffin

River Cam,
Cambridge

Where Now for Hawking?

A hectic schedule

Cambridge University custom dictates that the Lucasian Professor of Mathematics steps down at the age of **67**, and Hawking did so in 2009 after 30 years in the post. He continues his hectic schedule as a director of DAMTP, supervising students, researching, giving talks overseas, and writing popular-science books. Among these, *The Universe in a Nutshell* gave key updates on advances in theoretical physics and *The Grand Design* (co-written with **Leonard Mlodinow**) sparked controversy with its bold statements regarding religion and science. In 2010, the television series *Stephen Hawking's Universe* aired on the Discovery Channel offering viewers a popular account of modern astrophysics with subjects ranging from the nature of the Universe to the chances of alien life and time travel.

Science ambassador

Despite his physical disabilities, Hawking continues to be a great roving ambassador for theoretical physics. In 2008, he was awarded the first **Distinguished Research Chair** at Perimeter Institute (PI) in Ontario, Canada, one of the world's leading centers of theoretical physics. Hawking chose PI specifically because of its twin focus on quantum theory and gravity and has enjoyed regular extended stays there. PI honored Hawking by naming a new **state-of-the-art building** after him—and he was there for its grand opening in September 2011. The center doubles the size of the initial complex and accommodates up to 250 physicists, making **PI** the largest research facility in the world devoted solely to theoretical physics.

A medical milestone

In 1963, Hawking was told that he had only two years to live. On January 8, 2012, he turned **70**, making him one of the **longest-surviving** ALS sufferers in the world. He has often attributed his remarkable longevity to the support of his loving family and the focus that his work has provided. The accolades continue to flood in, and Hawking now has **12 honorary degrees**, a CBE (Commander of the Most Excellent Order of the British Empire), and is a Companion of Honour—a chivalric award bestowed on a person who has given conspicuous service to Britain. In 2009, Barack Obama awarded Hawking the Medal of Freedom—the highest civilian honor in the United States. People are fascinated by the contrast between Hawking's limited physical powers and the vast nature of the Universe he considers. For many he represents the ultimate example of mind over matter.

3-Second Brief
Hawking celebrates his 70th birthday—a remarkable physical achievement to add to a lifetime of academic success.

Related Thoughts
see also
LASTING LEGACY
page 150

❝I have a beautiful family, I am successful in my work, and I have written a bestseller. One really can't ask for more.**❞**

2001

Neil Turok and Paul Steinhardt go public with their "ekpyrotic theory," a string-based model describing the long-term behavior of the Universe.

2003

Hawking participates in an anti-war demonstration in London's Trafalgar Square where he calls the US-led invasion of Iraq a "war crime."

2003

Hawking shows his maverick side as he pays a visit to Stringfellow's lap-dancing club.

2006

A TV campaign starring Hawking wins awards for challenging perceptions regarding the disabled and boosting awareness of their needs.

2007

Gonville and Caius College, Cambridge, opens the Stephen Hawking Building—a center with purpose-built accommodation for disabled students.

2009

After Republicans in the US brand the UK's National Health Service "evil" and "Orwellian," Hawking defends it by saying that he would not be alive without it.

2010

The UK sees a much-needed surge in the number of students choosing to study mathematics and physics at school and university. Hawking, and his contemporaries, have made it "cool to be clever."

2010

In his TV series *Stephen Hawking's Universe*, Hawking warns that alien life forms may not be friendly.

2011

Saul Perlmutter, Adam Riess, and Brian Schmidt win a Nobel prize for their discovery of "dark energy"—thought to be the cause of the accelerating expansion of the modern Universe.

January 8, 2012

Hawking celebrates his 70th birthday. He has lived with ALS for almost 50 years to become one of the world's greatest theoretical physicists.

Glossary

Asteroid 7672 Hawking This minor planet named after Hawking was discovered in 1995 in the main belt of our Solar System, circling the Sun between Mars and Jupiter.

BAFTA British Academy of Film and Television Awards is the UK equivalent of the Oscars®. *Hawking*, a 2004 drama about his early years, received two nominations.

Causal sets An approach to quantum gravity that breaks spacetime down into a network of discrete points that can be linked without traveling faster than light.

Conformal cyclic cosmology Theory of the Universe put forward by Roger Penrose, based upon the idea of stitching together a (potentially infinite) number of ordinary Big Bang cosmologies.

Cuban missile crisis The closest the world has come to nuclear war, when, in 1962, the Soviet Union stationed bombers and medium-range nuclear missiles on Cuba.

Futurama Hawking has made a number of guest appearances in this sci-fi animated series from *The Simpsons* creator Matt Groening.

Gravitational waves A strong gravitating body—like a black hole or neutron star—that moves very rapidly generates ripples that travel outward through space and time, called gravitational waves.

Large Hadron Collider A giant machine on the French–Swiss border for slamming subatomic particles together at near lightspeed in order to better understand their physics. It is the world's most powerful particle accelerator.

Many worlds interpretation This theory says that events which happen with a given probability in our Universe will happen for definite in one of a network of parallel Universes disconnected from our own.

Multiply connected universe The idea that our Universe could be spherical or doughnut shaped—so that if you leave on one side you re-enter on the side opposite.

Possum Formerly Cambridge Adaptive Communications, which built the chair-mounted computer system used by Hawking. Such technology has progressed to allow disabled people greater independence at home and at work

Primordial black hole Black holes formed in the intense pressure of the early Universe. They weigh 10^{11} pounds but measure less than a thousand-billionth of one millimeter across.

Quantum information theory The study of the incredible properties of information—which could be a stream of computer data, made up of 1s and 0s—down in the quantum world.

Self-consistency Says that time travel can only alter the past in a way that is logically consistent with the future.

Supermassive black hole Gargantuan black holes thought to lie at the center of almost every galaxy in the Universe, including our own.

Supersymmetry Theory in particle physics which, broadly speaking, posits deep similarities between the two major types of elementary particles (known as bosons and fermions).

Stem cell research Hawking has criticized former president George W. Bush and the EU over their reluctance to develop stem cell research and which, he believes, holds the promise to treat disabilities such as ALS.

Toy model A highly simplified theory in science, which researchers sometimes build as a "proof of concept" or to simply "get their heads around" a problem, before attempting a full solution. An example might be the two-dimensional models of quantum black holes that were developed in the 1990s as a forerunner to full four-dimensional theories.

Trident system Britain's nuclear missile system, which is due for a major and extremely expensive overhaul by 2020. Hawking is a vocal opponent of nuclear weapons and has spoken out against the plans for Trident.

Vilenkin initial state An alternative to the Hartle–Hawking boundary condition for the Universe that limits spacetime to initial states that are expanding.

VoiceText A text-to-speech engine, from NeoSpeech, that produces a more realistic tone of voice. Hawking began using his "new voice" in summer 2011.

3-Minute Summary

Early life

Stephen William Hawking was born on **January 8, 1942**, in Oxford. The family moved to St. Albans, in the English county of Hertfordshire, when he was aged eight. Hawking's father was keen for his son to follow him into the medical profession. Hawking Jnr had other ideas, however, and went up to Oxford in 1959 to study mathematics and physics. Stephen was a **lazy student**—finding his degree course too easy—but nevertheless scraped through and secured a place at Cambridge to study for a **PhD in cosmology**. There, he produced a brilliant PhD thesis that established him as a rising star. In 1963, Hawking was diagnosed with **Amyotrophic Lateral Sclerosis** (ALS), an incurable wasting disease. He was helped through a period of despair by his work—and by a romance with a young language student named Jane Wilde, whom he married in 1965.

Family life

The Hawkings' first child, Robert, was born in 1967—followed by Lucy in 1970 and Timothy in 1979. Meanwhile, Hawking's work progressed in leaps and bounds, leading in 1974 to his **seminal discovery** that black holes emit a kind of radiation. Awards and accolades began to flood in, culminating with his appointment to the **Lucasian Chair of Mathematics** in 1979. In the 1980s, Hawking made contributions to quantum gravity research and inflation theory—and became involved in the search for a unifying "**theory of everything**." But there were storm clouds, too. In 1985, ALS-related pneumonia nearly claimed his life—only a tracheostomy saved him but at the cost of his voice. He was forced to use an **electronic voice synthesizer** instead.

Superstardom

The mounting cost of his care, and raising a family, led Hawking to write a popular-science book: **A Brief History of Time**. Published in 1988, it was an immediate bestseller, and propelled Hawking to global stardom. In 1990, he shocked the world by leaving Jane for his nurse, Elaine Mason—they married in 1995. The relationship eventually proved unsuccessful, however, and they divorced in 2006. Since the turn of the century, Hawking has continued to contribute a steady stream of research papers, dividing his time between Cambridge and, later, a **Distinguished Research Chair** at Canada's Perimeter Institute. In January 2012—the physicist who didn't expect to live past 23—celebrated his **70th birthday**.

> **"**I have lived with the prospect of an early death for the last 49 years. I'm not afraid of death, but I'm in no hurry to die. I have so much I want to do first. I regard the brain as a computer that will stop working when its components fail. There is no heaven or afterlife for broken-down computers; that is a fairy story for people afraid of the dark.**"**

Stephen Hawking, 1962

Stephen Hawking, 2007

Resources

Books

Black Holes and Baby Universes and Other Essays
Stephen Hawking
BANTAM BOOKS, 1994

Black Holes and Time Warps: Einstein's Outrageous Legacy
Kip Thorne
W.W. NORTON & COMPANY, 1995

The Black Hole War: My battle with Stephen Hawking to Make the World Safe for Quantum Mechanics
Leonard Susskind
BACK BAY BOOKS, 2009

A Brief History of Time
Stephen Hawking
BANTAM BOOKS, 1995

A Briefer History of Time
Stephen Hawking and Leonard Mlodinow
BANTAM BOOKS, 2008

The Elegant Universe: Superstrings, Hidden Dimensions and the Quest for the Ultimate Theory
Brian Greene
VINTAGE BOOKS, 2000

George's Secret Key to the Universe
Stephen and Lucy Hawking
SIMON & SCHUSTER, 2009

God Created the Integers
Stephen Hawking (ed.)
RUNNING PRESS, 2007

The Grand Design
Stephen Hawking and Leonard Mlodinow
BANTAM BOOKS, 2010

The Large Scale Structure of Space-Time
Stephen Hawking (et al)
CAMBRIDGE UNIVERSITY PRESS, 1975

On the Shoulders of Giants
Stephen Hawking
RUNNING PRESS, 2003

Stephen Hawking: A Life in Science
Michael White and John Gribbin
JOSEPH HENRY PRESS, 2002

Stephen Hawking: Quest for a Theory of Everything
Kitty Ferguson
BANTAM, 2003

The Universe in a Nutshell
Stephen Hawking
BANTAM BOOKS, 2001

Magazines/articles

10 Questions for Stephen Hawking
Time Magazine, November 15, 2010
www.time.com

How to Build a Time Machine
Daily Mail, April 27, 2010
www.dailymail.co.uk

Why God Did Not Create the Universe
Wall Street Journal, September 3, 2010
online.wsj.com

Life and the Cosmos, Word by Painstaking Word: A Conversation with Stephen Hawking
The New York Times, May 9, 2011
www.nytimes.com

Stephen Hawking: 'There is no heaven; it's a fairy story'
The Guardian, 15 May 2011
www.guardian.co.uk

Web sites

Hawking's personal web site
www.hawking.org.uk

Hawking's technical papers on Arxiv
arxiv.org/find/all/1/au:+Hawking/0/1/0/all/0/1

Hawking lecture transcripts
www.hawking.org.uk/index.php/lectures

Hawking's NASA 50th Anniversary Lecture
www.youtube.com/watch?v=3PCAGm5a1r8

Department of Applied Mathematics and Theoretical Physics
www.damtp.cam.ac.uk

Discovery Channel site for *Stephen Hawking's Universe*
dsc.discovery.com/tv/stephen-hawking/

Stephen Hawking's appearance in *Star Trek*
http://www.youtube.com/watch?v=mg8_cKxJZJY

MC Hawking spoof YouTube channel
www.youtube.com/user/therealmchawking/videos

Index

Acknowledgments

The publisher would like to thank the following individuals and organizations for their kind permission to reproduce the images in this book. Every effort has been made to acknowledge the pictures, however we apologize if there are any unintentional omissions.

Alamy/Miguel Sayago: 16B.
AllenMcC: 152, 117B.
Bantam Press: 125L.
Jacob D. Bekenstein: 117BR.
Nissim Benvenisty: 123TL, 123C.
Wayne Boucher/Cambridge2000.com: 149B.
Alex Brown: 57T.
Courtesy of the Clendening History of Medicine Library, University of Kansas Medical Center: 79C.
Martin Conway: 95T.
Corbis/Eleanor Bentall: 17; Bettmann: 67, 95C; Michael S. Yamashita: 113B.
David Dvir: 115.
Fotolia: 47TR, 147; Adimas: 47B, 61; Alex Fiodorov: 100, 101, 107R; Amorphis: 137TL; ArchMen: 101BR; Mademoiselle Bézier: 27; Kasia Biel: 103L; Christian Bijani: 89L, 94C, 95, 95BL, 107, 108; BiterBig: 37; Daniel Boiteau: 69C; Brilt: 99BR; Hugo C Campos: 94R, 151BR; Mardis Coers: 137BL; Sergii Denysov: 89; Marc Dietrich: 99BL; Karelin Dmitriy: 85BR; Christopher Dodge: 35T; Elenathewise: 54; Jürgen Fälchle: 83, 85C; Stephen Finn: 70; Paul Fleet: 81T; Foto Flare: 51; Frenta: 6, 99T; Sean Gladwell: 37B, 112TR; Victor Gmyria: 83TL; Andreas Haertle: 54; Innovari: 146, 147; iQoncept: 72, 106R; Vaclav Janousek: 91C; javarman: 23C; Jezper: 101BL, 107BL; Anja Kaiser: 120B; Iakov Kalinin: 87BL; kikkerdick: 73R; Vladislav Kochelaevs: 79BL; Georgios Kollidas: 40TR; Ralf Kraft: 95R; ktsdesign: 127; Jon Le-Bon: 50C; Le Do:

87, 142; Robert Lerich: 25BL; Liveshot: 91B; Matamu: 91T; David Mathieu: 49; Yuriy Mazur: 59; Mopic: 29; Nmid: 33; Dennis Oblander: 137C; Tyler Olson: 105BL; Konovalov Pavel: 24; Mariia Pazhyna: 50T; Peresanz: 141T; Anton Prado Photo: 76TR, 107L; Karen Roach: 51B; Robybret: 117B; Rolffimages: 87TR; Salavan: 55TR; Gina Sanders: 55; Gino Santa Maria: 55B; Dmitry Sosenushkin: 137CL; SSilver: 127T; James Steidl: 28, 36; David Talakhadze: 57B; Tauro79: 128T, 129T; Terex: 53T; Andrzej Tokarski: 25BR; Guillaume Tunzini: 43R; Tuulijumala: 53B; Vetkit: 137BR; Viktor: 126TR; Vincentmax: 93B; Viper: 97; Wawro Designs: 22; Ivonne Wierink: 137CR; Jan Will: 75R; Zagorskid: 95BR.
Biswarup Ganguly: 31B, 117L.
Getty Images/CBS Photo Archive: 103C; China Photos: 15T; David Montgomery: 35R, 39BL; WPA Pool: 15.
Private Collection of Gary Gibbons: 115.
Shai Gil: 59B, 61.
Dave Guttridge, The Photographic Unit: 115.
JD Hancock: 141C, 152.
Sean Hickin: 39TR.
Matthew Hunt: 23B.
David Iliff: 29B.
iStockphoto/Gustaf Brundin: 75, 93T, 143;
ChuckSchugPhotography: 147C.
Bob Lee: 131B.
Library of Congress, Washington, D.C.: 21C, 25B, 31T, 46, 48, 60, 69B, 79TL, 106L.
NASA: 17T, 142, 143; ESA, AURA/Caltech, Palomar Observatory: 101TL; ESA/Hubble: 100, 101TR; Hui Yang University of Illinois O: 2; Jeff Hester, and Paul Scowen (Arizona State University): 64; JPL: 6, 27; JPL-Caltech/R. Hurt: 95TR; Kim Shiflett: 55C, 123B; STScI: 139BC.
Keenan Pepper: 117BC.

Rex Features: 10, 11TL, 11TR, 11BL, 11BR, 21BR, 23T, 25, 29T, 33, 157; 20thC.Fox/Everett: 14B, 127C; Jonathan Hordle: 121; ITV: 14T; Nils Jorgensen: 53C, 157; Tony Kyriacou: 16T; Robin Laurance: 45B; Greg Mathieson: 125B, 154; Olycom SPA: 129B; Matthew Power: 17BL; Geoffrey Robinson: 151TL; Sipa: 15BL, 119T; Startraks Photo: 12B.
David Samuel: 125TL.
Maurice Savage/NASA/ The Royal Society: 57C.
Scala Archives/Scala Florence/Heritage Images: 13T.
Science Photo Library: 49L, 81L; Corbin O'Grady Studio: 131C; David Parker: 13B.
Shutterstock/Agsandrew: 145; AlessandroZocc: 127BC; Anteromite: 69, 115; Alin B: 125TR; Clearviewstock: 135T; 147; Corepics: 113T, 152; Pavel Gaja: 144, 145; Markus Gann: 4, 62, 77C, 105R, 107, 139T; Gataki: 103B; gingging: 105, 105; Patrick Hermans: 149T; debra hughes: 66, 93T, 109; Imagewell: 73C, 137T; Image Wizard: 6, 18, 132, 133R; Ladyann: 79TR; MarcelClemens: 149C; Inga Nielsen: 79; Nomad_Soul: 145; Oneo: 91B; OPIS: 73B; Picsfive: 83TR; Leigh Prather: 151BL; Christophe Rolland: 83BR; sdecoret: 135L; 152; Shrewsbury Design and Photography: 143; Snaprender: 74; Michael Taylor: 95BC; Clifton Thompson: 143; tovovan: 69B; trucic: 87BR; Vicente Bracelo Varona: 133C; Vladru:135; Vlue: 157; Mansurova Yulia: 85B.
Pete Souza: 59C.
Kevin Stanchfield: 43TL.
Hedwig Storch: 143.
Homer Sykes/ www.homersykes.com: cover, 37T.
Topfoto/John Hedgecoe: 12T.
Elke Wetzig: 121TL.
Wikipedia/Ojan: 131T.
Andy Wright: 119B, 153.